YOUR KNOWLEDGE HAS VALUE

- We will publish your bachelor's and master's thesis, essays and papers

- Your own eBook and book - sold worldwide in all relevant shops

- Earn money with each sale

Upload your text at www.GRIN.com and publish for free

Bibliographic information published by the German National Library:

The German National Library lists this publication in the National Bibliography; detailed bibliographic data are available on the Internet at http://dnb.dnb.de .

Imprint:

Copyright © 2018 GRIN Verlag
Print and binding: Books on Demand GmbH, Norderstedt Germany
ISBN: 9783668634909

This book at GRIN:

https://www.grin.com/document/385849

Afras Abera

Review on Chicken Production in Ethiopia with emphasis on Meat Production

GRIN Verlag

Review on Chicken Production in Ethiopia emphasis on Meat Production

By Afras Abera Alilo

Jimma University P. O.Box 307, Jimma Ethiopia

Abstract

This study Reviews on Chicken Production in Ethiopia emphasis on Meat Production with the aim of delivering summarized information for the beneficiaries and reader. Poultry production is increasing globally due to increase population number and increased demand of poultry products. Poultry production plays a great role for the supply of egg and meat from rural and urban area as a source of small holder family income and poultry production is synonymous with chicken production under the present Ethiopian conditions. In Ethiopia, poultry production systems show a clear distinction between the traditional, low input system on the one hand and modern production system using relatively advanced technology on the other hand. poultry production is very important by supplying nutritional need of humans, alleviating poverty, requiring small space and capital for investment; sources of family income and benefit to female and children and a lot of socio economic and cultural benefit and affected by different factors shortage an high cost of feed; disease and biosecurity and environmental constraints. Due to different constraints the meat production capacity of local or indigenous chicken is low; so it is difficult to supply the meat need humans by using indigenous breed because indigenous breeds are dual purpose with low production capacity and should be genetically upgraded by selecting superior birds and cross breeding with indigenous breeds to cope the low production capacity of indigenous.

Keywords*; chicken, Ethiopia and production*

1. Introduction

Poultry includes all domestic birds kept for the purpose of human food production (meat and eggs) such as chickens, turkeys, ducks, geese, ostrich, guinea fowl, doves and pigeons. In Ethiopia ostrich, ducks, guinea fowls, doves and pigeons are found in their natural habitat whereas, geese and turkey are uncommon. Thus poultry production is synonymous with chicken production under the present Ethiopian conditions (Solomon, 2007).

In Ethiopia chickens are the most widespread and almost every rural family owns chickens, which provide a valuable source of family protein and income (Tadelle *et al.*, 2003). The total chicken population in the country is estimated to be 56.5 million with native chicken representing 96.9%, hybrid chicken 0.54% and exotic breeds 2.56% (CSA, 2014).The most dominant chicken types reared in Ethiopia are local ecotypes, which show a large variation in body position, plumage colour, comb type and productivity (Halima, 2007). However, the economic contribution of the sector is not still proportional to the huge chicken numbers, attributed to the presence of many productions, reproduction and infrastructural constraints (Aberra, 2000; Halima, 2007)

The chicken production system in Ethiopia can be characterized by not market oriented, low input, scavenging and traditional management system consisting of local breeds (Alemu and Tadelle, 1997). The indigenous birds are small in body size and low producers of meat and egg (EARO, 2000). For example, the productivity of scavenging hens is 40-60 small-sized eggs/bird/year (Tadelle 1996; Alemu and Tadelle, 1997). The total chicken egg and meat production in Ethiopia is estimated to be about 78,000 and 72,300 metric tons, respectively (Tadelle, 1996).

Despite the large population and the great role of chicken both to the livelihood of resource poor farmers and to the national economy at large, the current level of on-farm productivity in the smallholder production system is low due to various factors such as biological, social, economic and institutional factors. On the other hand, the modern poultry production system is very small in size and confined to urban and peri-urban areas and contributes less than 2% of eggs and meat production in the country.

One and the major suggested solution to increase the production and productivity of chicken is utilization of chicken production system which is modern, market oriented and compatible with the existing situation of the farming system. Therefore, by minimizing the production constraints through use of selected and productive poultry breed as well as improvement of the production system (feeds and feeding, housing, health, etc), it is possible to supply chicken products for the market demand over the household consumptions.

Therefore, the objective is to review available chicken production system in Ethiopia more focusing on broilers (meat) production.

2.   Chicken production system

In Ethiopia, poultry production systems show a clear distinction between the traditional, low input system on the one hand and modern production system using relatively advanced technology on the other hand (Yami, 1995).

Chicken can be reared in different management and production systems. Based on chicken breed type, input and output level, mortality rate, type of producer, purpose of production, length of broodiness, growth rate and number of chicken reared. In Ethiopia, there are three types of chicken production systems (ANRS BoARD, 2006). These are free-range production system, semi-intensive production system and intensive production system.

2.1   Free-Range Chicken Production System

This chicken production system is practiced in most rural areas of the country and objectives of production are for household consumption and as source of additional income for the household. It covers 95-98% of the chicken production system of the country and it is not profitable since it is not market oriented. It contains small flock size (5-20 chickens per household) which is indigenous breed types mostly depend on locally available feed material as supplement with low health services and other management practices.

The chicken does not have their own constructed chicken house rather maintained in the main house with the family. Chicken brooding and rearing is only the care they obtain form their mother/hen. Because of these there is high mortality of chicken and long broody periods and there is risk of exposure for different chicken diseases and predators.

The major feed sources for chicken are worms obtained from free scavenging, legumes, and cereals and sometimes there is supplemental feed during feed shortage. The amount given is small and do not fulfill their nutrient requirement. Because of this their productivity is low.

Advantages of free-range chicken production system

The advantages of free-range chicken production system include, the chickens are healthy since they exercise in the open air freely, there is minimal infection with parasites if enough space is available, there is little or no labour input, the chickens in this type of production system help to limit the amount of rubbish in a productive way and the direct costs of the system are low.

Disadvantages of free-range chicken production system

The disadvantages of free-range chicken production system include, it is difficult to control and manage the chicken especially the young chicks are easily exposed for predators and unfavorable weather conditions, the chickens eat sown seed when looking for feed, a large percentage of the eggs can be lost as the laying hens are not familiar to laying nests, high diseases transmission and occurrence of high death, chickens are less productive.

2.2   Semi- Intensive Chicken Production System

This type of chicken production system is better than free ranging production system since it uses inputs like supplemental feed, vaccine, etc. It has a small house which accommodate laying nest and feeders which serves as chicken house for night time. The house has one or two side open door for easy movement of the chicken to the fenced area during the day time. The fence can be made from mesh wire or other materials and will not allow the chicken to escape above on it. The fenced area should be always clean and dry. Since the feed the chickens obtain from the scavenging is very low, they should be supplemented with energy and protein feeds. Since the main objective of the production is to get profit, they should get better health management practice like vaccination against NCD than free scavenging system. They are more productive than the chicken in free scavenging system. It contains flock size of 50-200 birds/chicken per household which are improved breeds.

Advantages of semi-intensive chicken production system

The advantage of this system include, complete control over operation, useful for record purposes, operational throughout the year, economic use of land (free range) and there is better protection during winter.

Disadvantages of semi-intensive chicken production system

The disadvantage of this system of chicken production system include, high cost in fencing, danger of over stocking and exposure for different disease if the site is not clean and dry.

2.3   Intensive Chicken Production System

This type of chicken production system use more inputs (feeds and feeding, breed, health, housing and other inputs) than the above two chicken production systems. It is market oriented and the main objective of production is to get better profit. The number of chickens involved are relatively high (more than 200 chicken). The chicken breed used is specialized improved breeds (layer or broiler). They should provide the expected product within that time.

There are few private large scale commercial poultry farms, all of which are located in Debre -Zeit. ELFORA, Alema and Genesis are the top 3 largest commercial poultry farms with modern production and processing facilities. ELFORA annually delivers (www.ethiomarket.com elfora), around 420,000 chickens and over 34 million eggs to the market of Addis Ababa. Alema poultry farms is the 2nd largest commercial poultry farms in the country delivering nearly half a million broilers to Addis Ababa market each year. The large scale commercial poultry Provide fertile eggs, table eggs, day old chicks, broiler meat and adult breeding stocks to the small scale modern poultry farms. They are kept as full time business and highly dependent on market for inputs. The general indications are that the intensive poultry industry plays a key role in supplying poultry meat and eggs to urban markets at a competitive price. The industry also provides employment for a range of workers from poultry attendants to truck drivers to professional managers.

Formal marketing operations exist in urban and peri-urban areas practicing large scale commercial poultry production. The majority of the products sold within the formal sector come from the commercial industry but a small number of frozen indigenous chickens are supplied through supermarkets in Addis Ababa. The larger commercial poultry units have agreements with their clients and most poultry meat is sold frozen. The majority of the products sold within the formal sector come from the commercial industry but a small number of frozen indigenous chickens are supplied through supermarkets in Addis Ababa. Dressed poultry carcass and table eggs are sold to residents and hotels either in supermarket or small shops/kiosks. Most of the supply of dressed poultry carcass to Addis Ababa supermarket came from Alema poultry farm but many unidentified sources also supply the supermarkets.

There are three types of intensive chicken production systems which include deep litter system, cage system and full slated rearing which are stated bellows in detail.

2.3.1 Deep Litter System

It involves rearing of chickens on a floor littered by 5-10 cm thickness litter. The litter can be made from locally available material such as dry hay, teff straw, coffee pulp and sow dust. The litter should be dry at any time otherwise it can cause occurrence of different disease.

In addition to provision of comfort for the chicken, the litter absorbs any waste material excreted from the chicken and makes the house dry. It is possible to place the feeders and drinkers in the house on the litter. But it is advisable to hang them as the age of the chickens increase. It is also important to place laying nest in the house. In this way it is possible to rear either layer or broiler. At least once a week, the litter should be sprayed with disinfectant chemicals. Deep litter is recommended for both meat birds and layers.

To keep healthy chicken in this type of system, the following points should be noted:

- The existing litter should be removed totally when the existing stock culled
- Before introduction of new stock, the house should be cleaned carefully and left free at least for two weeks
- Make sure that the litter should be dry at any time
- At any time the thickness of the litter should not be less than 5cm

The advantages of the system included proper accommodation, prompt culling of unproductive birds, proper control of diseases and predators, good record keeping and high egg production. It has also advantageous in that land requirement is minimum, easy and economic management, scientific feeding and management, high degree of supervision, minimum labour, automation is possible and manure value is increased.

Deep litter is a good insulation. It protects chickens from cold weather, and during hot seasons they can nestle into it and reach the cool floor below. Studies show that when all other factors are equal, layers produce more eggs on deep litter than in cage systems. Chickens can be brooded and kept through their productive lives in the same house. Deep litter allows the bird to dust itself against lice and other parasites.

The disadvantages of this intensive system of chicken production include high capital investment, problem of cannibalism and diseases outbreak. If the management is bad, liberation and accumulation of ammonia, wet litter problem dirty eggs, and disease problems may result. There is a greater chance of worm and tick infestation and coccidiosis (internal protozoan parasites) than with cages or raised floor systems. The deep-litter system is inappropriate for very humid areas (80 - 90% humidity) damp litter spreads diseases. The litter must be turned often, particularly in damp weather, and this requires more labor than other systems. Sometimes adequate litter is difficult to obtain.

2.3.2 Cage System

This type of intensive production system involves rearing of chicken in one house on the prepared cages or nests and it is more appropriate for layers. The chicken has no any access for free ranging. Therefore, the chicken should get the required nutrient through supplementation. The ration can be formulated in the house using locally available materials like maize, noug seed cake and other materials. Even though the system requires high initial capital investment, it is profitable. The rearing cage can be made from locally available materials like timber and wood. Cages are good for climates with high humidity, where labour costs are high, and when a farmer wants to keep a large flock of layers. Where ticks are a problem, cages are especially advantageous. Cages are recommended for layers, but not generally used for meat birds.

The advantages of the system include cages can be placed under existing roofs; thus, a special building may not be required. With cages more birds can be kept in a building than on deep litter. Less labour per bird is needed than other systems. Poor layers can be identified immediately and culled, thus saving feed.

Problems with parasites, particularly ticks, are reduced, but nutrition may be a problem. When properly constructed, cages can last many years. Fewer disease problems are caused by transmission through fecal matter. Cages are a cheaper investment in the long run due to ease in care and feeding of the birds

The disadvantages of the system include high cost of installation, breeding is not possible unless artificial insemination is practiced, cage layer fatigue or paralysis is a problem if not attended to, cages are hard to construct properly, and they involve very high initial investment per bird. There must be necessary vitamins constant and excellent ventilation. There are more broken eggs than with deep litter. The feed must contain all and minerals needed by birds.

Table 1 the major characteristics of the chicken production system in Africa

Characteristics	Traditional free range	Backyard or Subsistence	Semi-intensive	intensive
Flock Size	1-10	10-50	50-500	500-10000
Ownership	Women & children	Women & family	Middlemen	Business men
Breeds	Indigenous	Indigenous & crossbreeds	Cross breeds	Layers or broilers
Feed Sources	Scavenging	Scavenging & supplements	Commercial/local	Balanced diets
Health Status	No vaccination No medication	Vaccination & Little medication	Vaccination and Little medication	Full vaccination Full medication
Housing	No specific housing	Simple and small houses	Medium & improved	Big and improved
production	low	low	medium	high

Source: Sonaiya, E.B. 1990; Kitalyi, 1998; Sonaiya *et al.*, 1999; Gueye, 2003 and Riise *et al.*, 2004.

3. Major Constraints of Poultry Production

3.1 Inadequate Health Care and Poor Feed Source

The major problem impairing the existing production system in Ethiopia is the high incidence of Newcastle disease, which named locally "fengel" (Holye, 1992; Alemu and Tadelle, 1997; Solomon 2004). Another report in Southern parts of the country by Aberra (2007) indicated that fowl cholera is a major problem followed by Newcastle disease. Next to disease, the major limiting factor of production increase is lack of feed. The nutritional status of local laying hens from chemical analysis of crop contents indicated that protein was below the requirement for optimum egg production the deficiency is more series during the short rainy season and dry seasons (Tegene, 1992; Alemu and Tadelle, 1997)

3.2 Inadequate Emphasis To Research and Extension

Until recently, little emphasis has given to livestock and poultry research. The extension linkage between the research output and the ministry of agriculture and the farmers are found to be extremely weak (Alemu and Tadelle, 1997) and in general there is no consistent feedback to the research. Most of the poultry extension workers transfer their extension packages to the households expecting that the husband will pass the message to his wife (Fikre, 2000). However, poultry keeping in most parts of Ethiopia is mainly the responsibility of women as reported by Tadelle and Ogle (1996a).This indicated that there are no client oriented extensions.

3.3   Lack of Organized Market and Poor Access to Main Market

Even though chicken meat is relatively cheap and affordable source of animal protein (Alemu and Tadelle, 1997), lack of organized marketing system and the seasonal fluctuation of price are the main constraints of the poultry market in Ethiopia. Variation in price mainly attributed to high demand for chickens for Ethiopian New Year and holidays. It also partly influenced by weight, age of chickens and availability. The plumage color, sex, combs types, feather covers are also very important for influencing price. According to Gausi et al. (2004) the major constraints in rural chicken marketing were identified as low price, low marketable output and long distance to reliable markets. As a result, the smallholder farmers are not in a position to get the expected return from the sale of chickens. Likewise, poor marketing information system, poor access to terminal market, high price fluctuation and exchange based on plumage color, age and sex are among the main constraints of chicken market in the country (Kena, 2002)

3.4   Social and Cultural Constraints.

The socio cultural constraints to poultry development are the value placed up on poultry for use at ceremonies and festivals or even as source of income in times of need but neither as source of daily food nor as regular source of income. Some regard chickens as their pets or part of the family, thus rarely used as food for home consumption, although they can sold without regret and the money utilized. Another constraint is the social norm that determines owner ship of livestock. Typically, where crop farming is the men's main activity, keeping livestock is perceived as a peripheral activity neglected to women and children. Practical experience indicates that there were no regular watering and supplementing feed and they do not clean the birds' night shelter and take care of the young chicks. Farmers are also reluctant to expand their poultry farm. The farmers attitude to the sector makes the rural traditional poultry farming remain unchanged for a long time.

3.5 Breed Constraints

A breed of the poultry is the main factor that is considered in chicken meat or broilers production. The meat production ability of indigenous chicken was limited in growth performance Bog ale (2008) Day old chickens of different populations of indigenous chicken measures live weight of 27.3g per chicken (Halima, 2007and Bog ale, 2008).

Nigussie (2011) in adult live body weight of the different populations of indigenous local chickens also reported 1.6 kg for male and 1.3 kg for females. According to Solomon, (2003) reported that there was no difference between White Leghorn and indigenous chickens raised under scavenging condition in mean daily body weight gain at 2 months of age. He also reported that the indigenous chickens are sold for meat purpose starting from 6-8 months of age at weight of around 0.7-1.4kg.

4. Chicken Meat Consumption in Ethiopia

Occasional consumption of eggs and poultry meat provides a valuable source of protein in the diet. For the poor, poultry meat is the only special meal they can afford during religious festivities like New Year, Christmas and Easter. In general, socio-cultural roles are more important in the area with the poorest market access (Aklilu, 2007). Unfortunately however, eggs have never been among the top ten animal products (milk, butter, cheese, honey, beef, mutton and goat meat etc.) consumed at the household level in rural areas of SNNP Regional State.

The national poultry meat consumption is estimated, on an average to be 69,000 tons per annum (ILRI, 2000). In the mid-1990s; the per capita egg and poultry meat consumption in Ethiopia was estimated at 57 eggs and about 2.85 kg meat, respectively (Alemu and Tadelle, 1997). However, the per capital annual poultry meat and egg consumption has been declining and estimated at the national average of close to 0.12 and 0.14 kg, respectively (USAID., 2006, 2010).

Table 2. Summary of population and per capital meat consumption in urban and rural area of Ethiopia (source data obtained from CSA 1996; 2000 and 2004)

residence	year	Population in millions of numbers	Per capita meat (kg/capita/year
rural	1996	7.6	7.8
	2000	7.6	7.6
	2004	9.1	9.7
urban	1996	45.1	1.7
	2000	48.4	2.7
	2004	55.3	2.6
country	1996	52.7	2.7
	2000	56.0	3.3
	2004	64.4	3.5

5.   Conclusion

Poultry includes all domestic birds kept for the purpose of human food production (meat and eggs) such as chickens, turkeys, ducks, geese, ostrich, guinea fowl, doves and pigeons. In Ethiopia ostrich, ducks, guinea fowls, doves and pigeons are found in their natural habitat whereas, geese and turkey are uncommon. Thus poultry production is synonymous with chicken production under the present Ethiopian conditions. In Ethiopia, poultry production systems show a clear distinction between the traditional, low input system on the one hand and modern production system using relatively advanced technology on the other hand. poultry production is very important by supplying nutritional need of humans, alleviating poverty, requiring small space and capital for investment; sources of family income and benefit to female and children and a lot of socio economic and cultural benefit and affected by different factors shortage an high cost of feed; disease and biosecurity and environmental constraints. Due to different constraints the meat production capacity of local or indigenous chicken is low; so it difficult to supply the meat need humans by using indigenous breed because indigenous breeds are dual purpose with low production capacity and should be genetically upgraded by selecting superior birds and cross breeding with indigenous breeds to cope the low production capacity of indigenous.

6.   References

Abera, M. 2000. Comparative studies on performance and physiological responses of Ethiopian indigenous (Angete Melata) chickens and their f_1 crosses to long term heat exposure. PhD dissertation, Martin-Luther University. Halle-Wittenberg Germany. pp127.

Aberra, M. and Tegene, N. 2007. Study on the characterization of local chicken in Southern Ethiopia.

Aklilu .H.M., 2007.Village poultry in Ethiopia; socio-technical analysis and learning with farmers. PhD Thesis, Wageningen University, Wageningen, the Netherlands Yami, A., 1995. Poultry production in Ethiopia. World's Poult. Sci. J., 51: 197-201.

Alemu, Y. and Tadelle, D. 1997. The status of poultry research and development in Ethiopia, research bulletin No.4, poultry commodity research program Debre Zeit agricultural research center. Alemaya University of agriculture, Ethiopia

Bogale, K., 2008. In situ characterization of local chicken eco-type for functional traits and production system in Fogera district, Amahara regional state. M.Sc. Thesis submitted to the department of animal science school of graduate studies, Haramaya University, pp: 107

CSA [Central Statistical Agency- Ethiopia (2014/15[2007 E.C.]): Agricultural sample survey: Livestock and livestock characteristics

Fikre, A. 2000. Base line data on chicken population, productivity, husbandry, feeding and constraints in four peasant associations in Ambo Wereda. Department of Animal Sciences, Ambo College of Agriculture, Ambo, Ethiopia.

Gausi, A., Safalaoh, J., Banda, D. and Ongola, N. 2004. Characterization of small holder poultry marketing systems in rural Malingunde: a case study of Malingunde extension planning area; Nt Chell University of Malawi, Bunda College of Agriculture, Lion We, Malawi

Gueye, E.F. 2003. Poverty alleviation, food security and the well-being of the human population through family poultry in low income food-deficit countries. Senegalese Institute of Agricultural research (ISRA), B.P.2057, Dakar-hann, Senegal.

Halima HM(2007) <u>Phonotypic and genetic characterization of indigenous chicken populations in Northwest Ethiopia.</u>

highlands of Ethiopia. M.Sc Thesis, Swedish University of Agricultural Sciences, Sweden.

Hoyle, E. 1992. Small-scale poultry keeping in Welaita, North Omo region. Technical pumpblet No. 3 Farmers Research Project (FRP). Farm Africa Addis Ababa

ILRI (International Livestock Research Institute). 2000. Handbook of livestock statistics for developing countries. Socio-economic and Policy Research Working Paper 26. ILRI, Nairobi, Kenya. 299 pp.

Kena, Y., Legesse, D., and Alemu, Y. 2002. Poultry marketing: structure, spatial variations and determinants of prices in Eastern Shewa zone, Ethiopia. Ethiopian Agricultural Research Organization, Debrezeit Research Center.

Kitalyi, A. J. 1998. Village chicken production systems in rural Africa household food security and gender issues: FAO, Rome.

Nigussie, D., 2011. Breeding programs for indigenous chicken in Ethiopia, Analysis of diversity in production systems and chicken populations. PhD. Thesis submitted in fulfillment of the requirements for the degree of doctor at Wageningen University Netherlands, pp: 148

Riise, J.C., Permin, A., Mc Ainsh, CV. and Frederiksen, L. 2004. Keeping village poultry. A technical manual on small-scale poultry production. Network for small holder poultry development.

Solomon D (2007) <u>Suitability of hay-box brooding technology to rural household poultry production system. Jimma University College of Agriculture and Veterinary Medicine, Jimma, Ethiopia.</u>

Solomon, D. 2004. Egg production performance of local and white leghorn hens under intensive and rural household conditions in Ethiopia. Jimma College of agriculture p.obox.307,Jimma, Ethiopia.

Sonaiya, E.B. 1990a. The context and prospects for development of smallholder rural poultry production in Africa. in proceedings, CTA seminar on smallholder rural poultry production, Thessaloniki, Greece, 9–13 October 1990, Vol. 1, p. 35–52.

Sonaiya, E.B., Branckaert, R.D.S., Gueye, E.F.1999. The scope and effect of family poultry research and development. Research and development options for family poultry. First-INFPD/FAO Electronic Conference on Family Poultry.

Tadelle Dessie and B.Ogle 1996. Studies on village poultry production systems in the central

Tegene, N. 1992. Dietary status of small holder local chicken in Leku, Southern Ethiopia. Sinet: Ethiopian journal of Science.

USAID., 2006. Partnership for Safe Poultry in Kenya (PSPK) program value chain analysis of poultry in Ethiopia. Winrock International, United States Agency for International Development (USAID), Ethiopia, pp: 1-42. http://pdf.usaid.gov/pdf_docs/PNADU075.pdf

YOUR KNOWLEDGE HAS VALUE

- We will publish your bachelor's and
  master's thesis, essays and papers

- Your own eBook and book -
  sold worldwide in all relevant shops

- Earn money with each sale

Upload your text at www.GRIN.com
and publish for free